To my STEAM parents, Tim and Martha —B.A.

To my wife, Rikako, and my children, Takeru, Kaory, and Nozomi —T.S.

To Mr. Darcy, the love of my life. —J.P.

Room to Read would like to thank Tatcha™ for their generous support of the STEAM-Powered Careers collection.

TATCHA

Copyright 2022 Room to Read. All rights reserved.

Written by Brooke Allyn
Featured scientist: Dr. Takeshi Saito
Illustrated by Janet Pagliuca
Edited by Jamie Leigh Real
Photo research by Kris Durán
Series art direction and design by Christy Hale
Series edited by Carol Burrell, Jamie Leigh Real, Jocelyn Argueta, and Deborah Davis
Copyedited by: Debra Deford-Minerva and Danielle Sunshine

ISBN 978-1-63845-061-0

Manufactured in the Canada.

MIX
Paper | Supporting responsible forestry
FSC® C011825

10 9 8 7 6 5 4 3 2
Room to Read
465 California Street #1000
San Francisco, California 94104
roomtoread.org

Room to Read

World change starts with Educated children.©

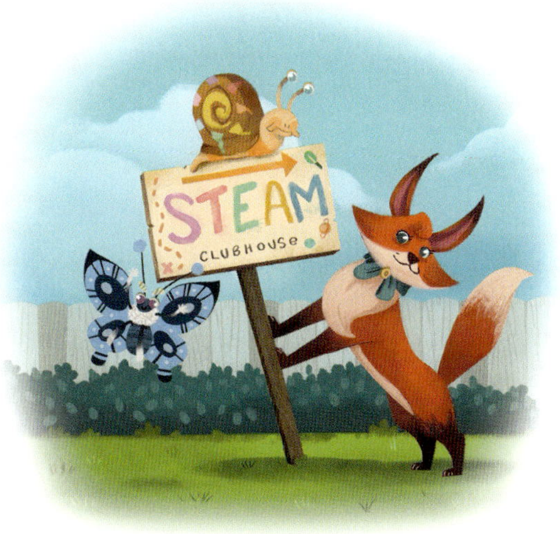
STEAM
CLUBHOUSE

STEAM-Powered Careers

GASTROENTEROLOGY

by **Brooke Allyn**

featured scientist: **Dr. Takeshi Saito**

illustrated by **Janet Pagliuca**

Room to Read

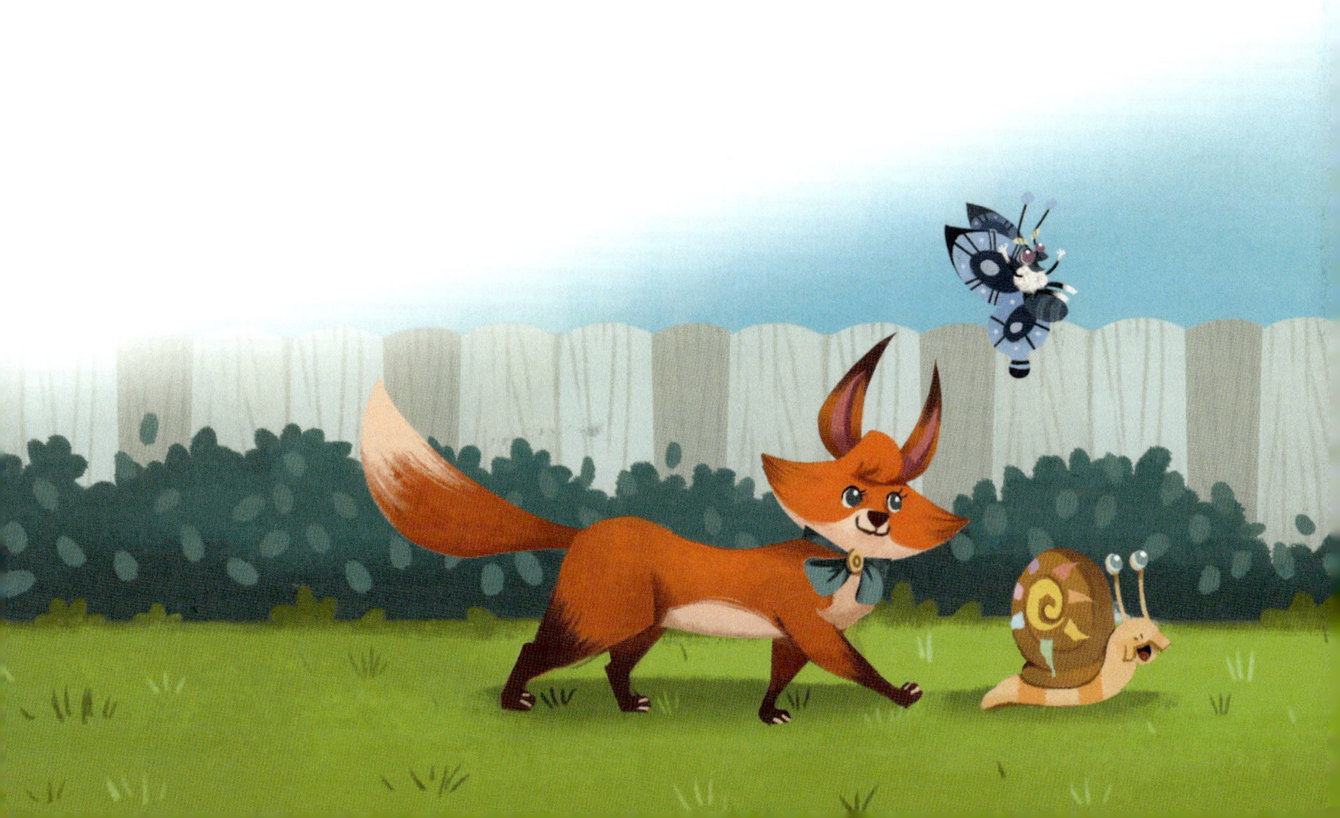

Contents

Explore Gastroenterology with Jae and Felicia	6
What Is Gastroenterology?	22
Meet the Scientist	24
Learn More about Gastroenterology	30
Word List	34

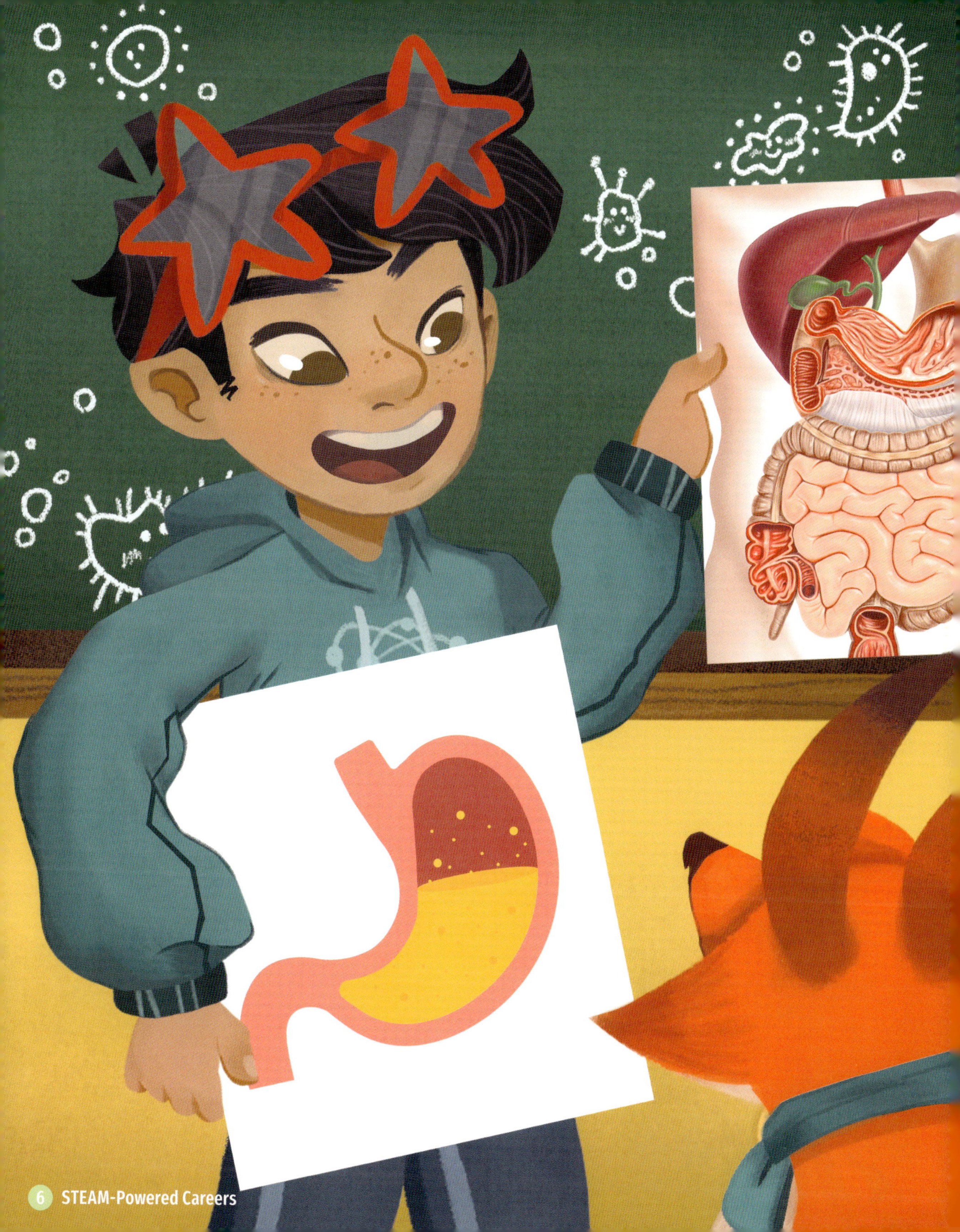

6 STEAM-Powered Careers

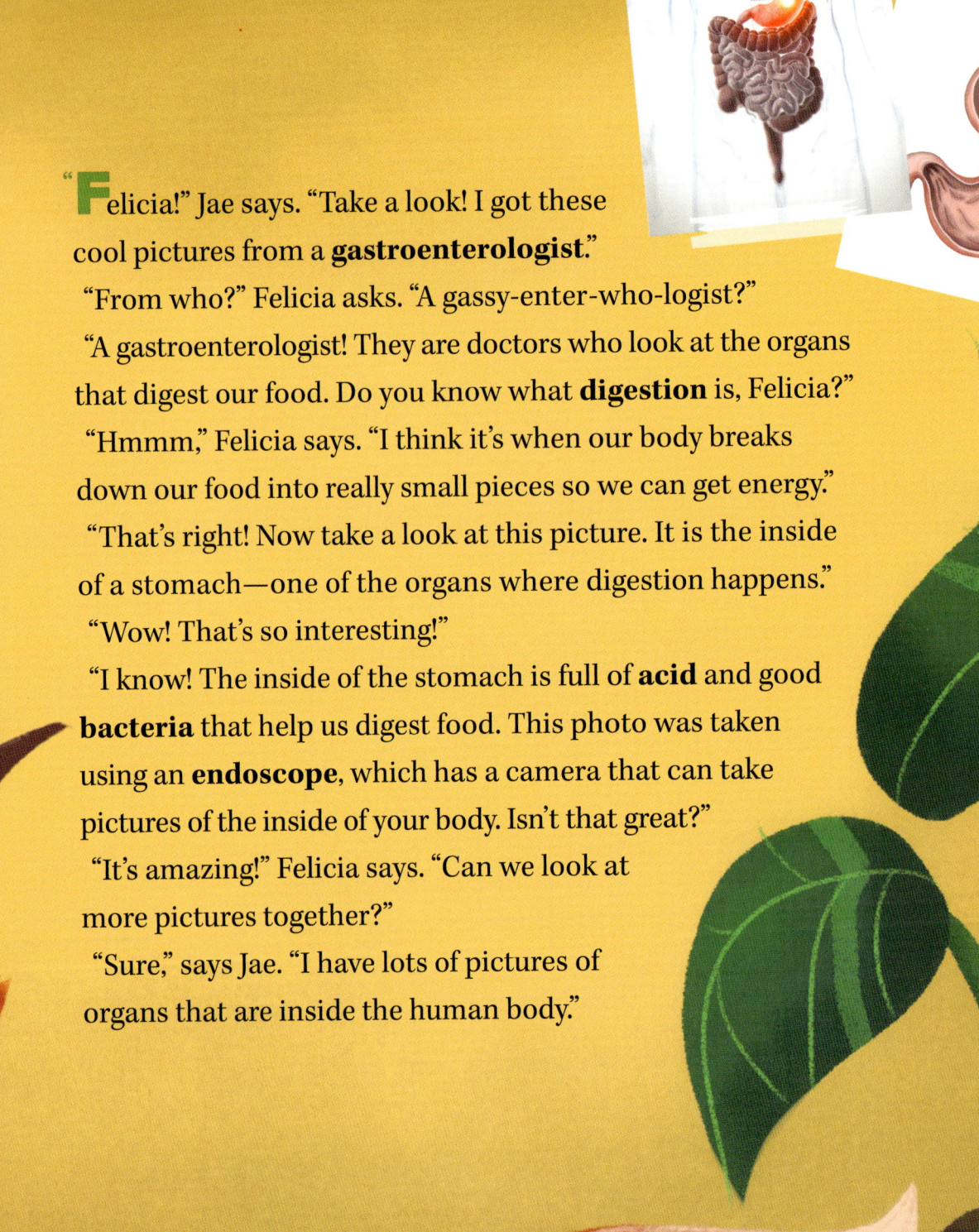

"**F**elicia!" Jae says. "Take a look! I got these cool pictures from a **gastroenterologist**."

"From who?" Felicia asks. "A gassy-enter-who-logist?"

"A gastroenterologist! They are doctors who look at the organs that digest our food. Do you know what **digestion** is, Felicia?"

"Hmmm," Felicia says. "I think it's when our body breaks down our food into really small pieces so we can get energy."

"That's right! Now take a look at this picture. It is the inside of a stomach—one of the organs where digestion happens."

"Wow! That's so interesting!"

"I know! The inside of the stomach is full of **acid** and good **bacteria** that help us digest food. This photo was taken using an **endoscope**, which has a camera that can take pictures of the inside of your body. Isn't that great?"

"It's amazing!" Felicia says. "Can we look at more pictures together?"

"Sure," says Jae. "I have lots of pictures of organs that are inside the human body."

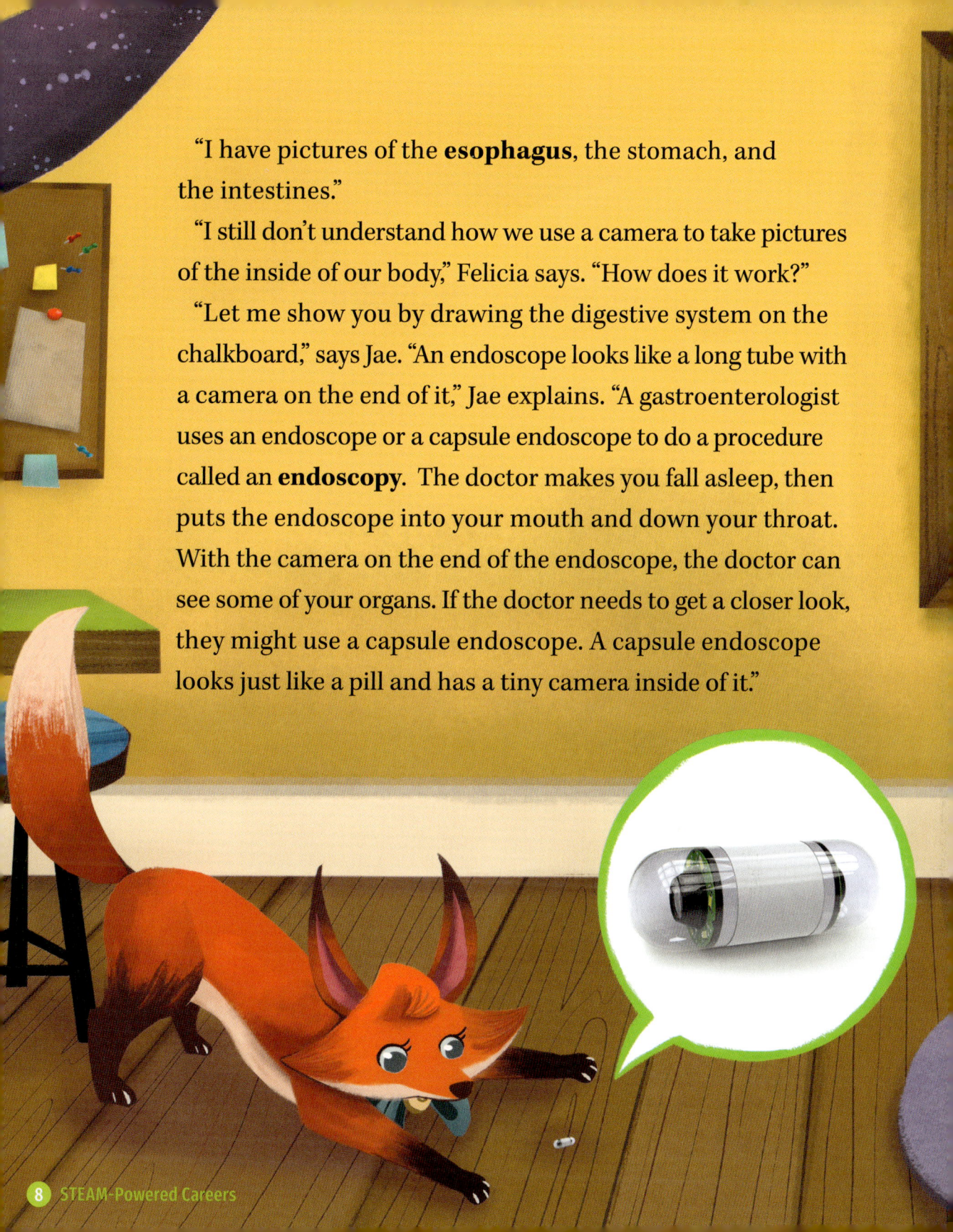

"I have pictures of the **esophagus**, the stomach, and the intestines."

"I still don't understand how we use a camera to take pictures of the inside of our body," Felicia says. "How does it work?"

"Let me show you by drawing the digestive system on the chalkboard," says Jae. "An endoscope looks like a long tube with a camera on the end of it," Jae explains. "A gastroenterologist uses an endoscope or a capsule endoscope to do a procedure called an **endoscopy**. The doctor makes you fall asleep, then puts the endoscope into your mouth and down your throat. With the camera on the end of the endoscope, the doctor can see some of your organs. If the doctor needs to get a closer look, they might use a capsule endoscope. A capsule endoscope looks just like a pill and has a tiny camera inside of it."

"Awesome!" Felicia says. "What's it like when the capsule goes through the body?"

"Let's find out!" Jae says, "We'll shrink ourselves and go on an adventure into the human body!"

Gastroenterology

WHOOSH!

Felicia and Jae shrink until they're really tiny. They climb inside a capsule endoscope—also called a pill submarine.

"Ready for our journey through the human body, Felicia?"

"Ready!"

"Okay! Our first stop: the mouth!"

"Whoa! What's all this sticky stuff?" Felicia says.

"That is saliva," Jae explains. "When you chew, your teeth and saliva work together to start digesting food."

"What happens after the food pieces are chewed up?" Felicia asks.

"You swallow them and they go down your esophagus," Jae says. "The esophagus is in your throat. Ready? It's a long way down."

Felicia looks down into the esophagus. "I'm a little scared. Will you hold my paw?"

"Of course! I'll always be here for my best friend."

"Do you feel the capsule bumping back and forth?" Jae asks. "The esophagus squishes and squeezes the food until it reaches the stomach."

Felicia giggles. "I feel it! What a bumpy ride, just like a roller coaster. This is really fun!"

"Up next: the stomach!"

The average adult esophagus is ten inches long.

There are more than six billion bacteria living in your mouth.

Gastroenterology 11

SPLASH!

Jae and Felicia land in a pool of smelly, clumpy liquid.

"Wow!" Felicia says, jumping around. "Look at this place! What happens here?"

"When food lands in the stomach, the stomach moves it around, breaking it up into even smaller pieces. This is called **mechanical digestion**. The stomach also mixes in bacteria, **enzymes**, and **stomach acid** that further break down the food. This is called **chemical digestion**."

When you eat food, you also swallow air. If your stomach has too much air in it, you burp to get it out.

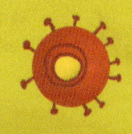

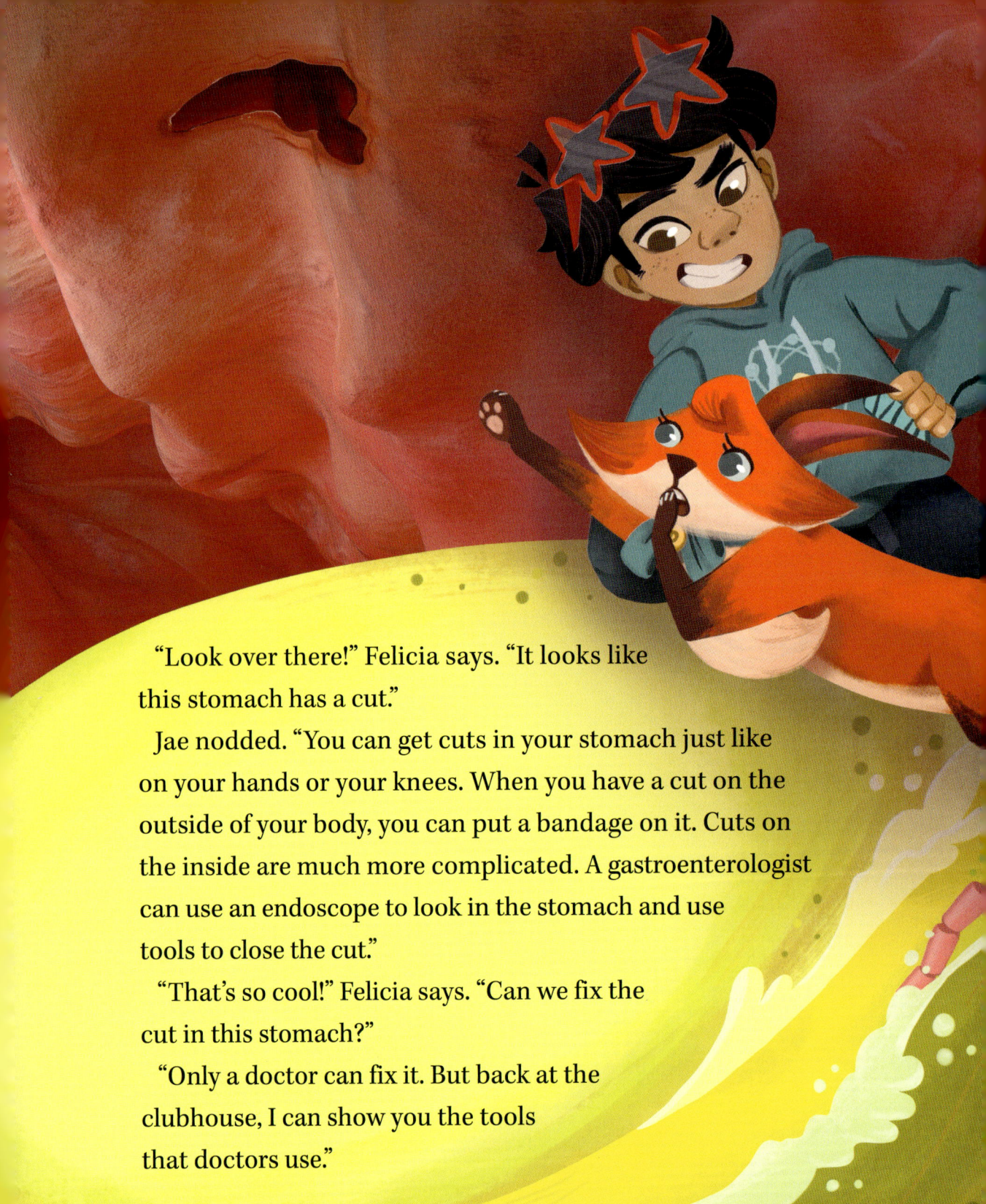

"Look over there!" Felicia says. "It looks like this stomach has a cut."

Jae nodded. "You can get cuts in your stomach just like on your hands or your knees. When you have a cut on the outside of your body, you can put a bandage on it. Cuts on the inside are much more complicated. A gastroenterologist can use an endoscope to look in the stomach and use tools to close the cut."

"That's so cool!" Felicia says. "Can we fix the cut in this stomach?"

"Only a doctor can fix it. But back at the clubhouse, I can show you the tools that doctors use."

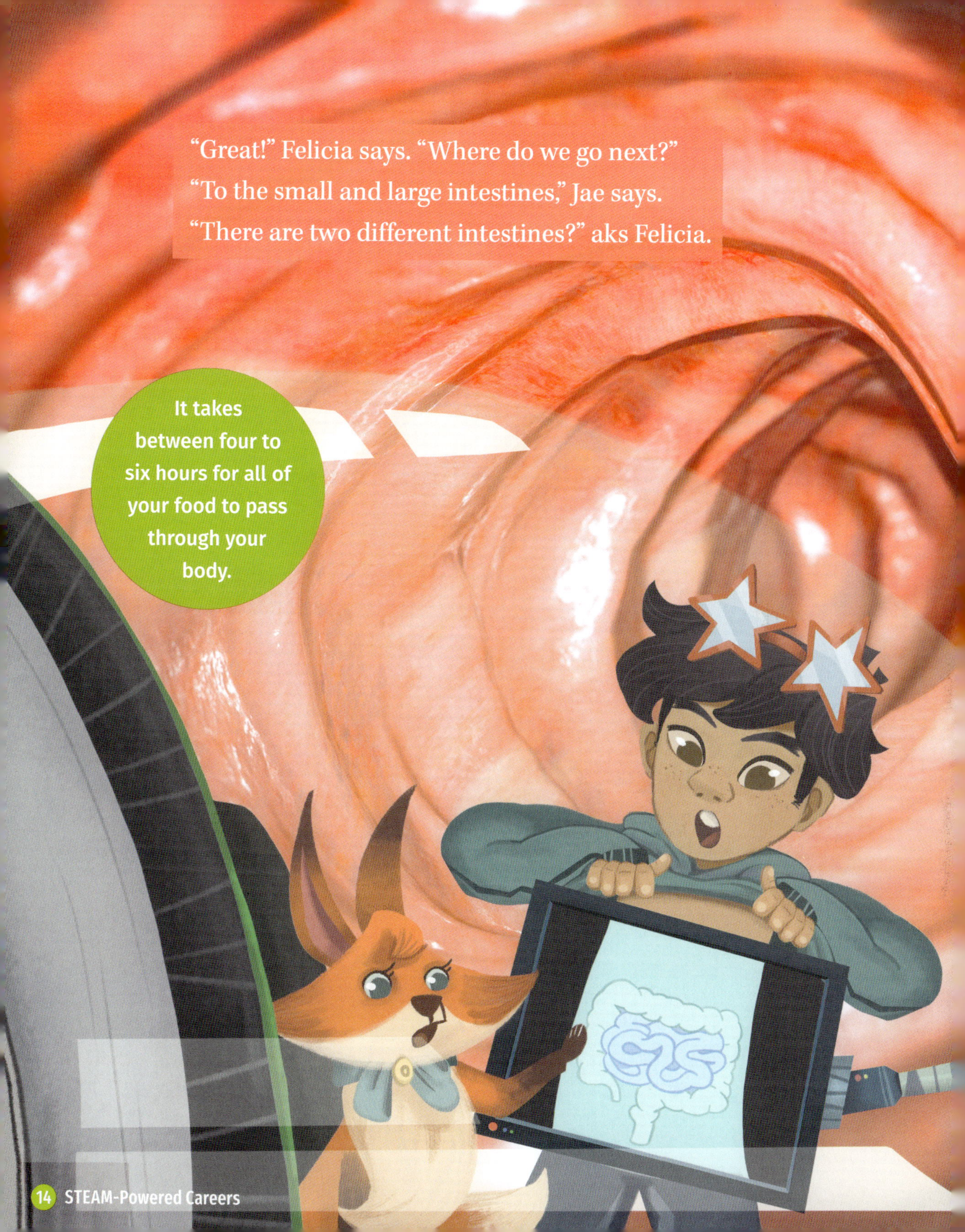

"Great!" Felicia says. "Where do we go next?"
"To the small and large intestines," Jae says.
"There are two different intestines?" aks Felicia.

It takes between four to six hours for all of your food to pass through your body.

14 STEAM-Powered Careers

"Yes!" Jae says. "First we'll go through the small intestine. It's full of tiny folds called **villi**. The job of the villi is to absorb **nutrients** from the food we eat. The nutrients give us energy and help our bodies grow."

"Those villi look so strange," Felicia says. "Like lots of little fingers." She wiggles her paws at him.

Jae laughs. "They sure do! Are you ready for the large intestine?"

"Whoa, Jae! Why is this ride so bumpy?"

"Not all of the food we eat can be used by our body. The waste that gets left behind moves on to the large intestine, where we are now. The large intestine uses the same movements that we felt in the esophagus and small intestine to move the waste along."

"This place is neat!" Felicia says. "The large intestine seems very strong. It's also very pink."

"A large pink intestine is a healthy intestine. A healthy intestine absorbs all the water from our food so our whole body can stay healthy. Then, whatever food waste is left over turns into poop."

"Poop?" Felicia yelps.

"Yes!" Jae says. "And now let's talk all about poop!"

The average adult large intestine is about five feet long.

Gastroenterology

"What do I need to know about poop?" Felicia asks. "I already know it's smelly."

"It's more than just smelly!" Jae says. "Poop is really important for the body. Did you know that a gastroenterologist can examine your poop to see how healthy you are?"

"How do they do that?"

"They check to see if the poop is too hard or too soft. They also check it for strange colors like green or yellow."

Poop can be green, brown, black, red, white, or yellow, but brown poop is the best color to have.

TYPES of POOP

1. Separate hard lumps *(constipation)*
2. Sausage-shaped but lumpy *(mild constipation)*
3. Sausage-shaped with cracks on its surface *(healthy)*
4. Smooth and soft like a snake *(healthy)*
5. Soft blobs with clear-cut edges *(lacking fiber)*
6. Fluffy and mushy with ragged edges *(mild diarrhea)*
7. Entirely liquid, watery *(diarrhea)*

STEAM-Powered Careers

"If your poop is yellow, for example, a gastroenterologist might say that your liver is sick."

"Is there a way to make someone's poop healthier?" Felicia asks.

"There are a few ways," Jae says. "One really neat way is called a poop transplant. A gastroenterologist gives a sick patient a pill filled with someone else's healthy poop bacteria. This good bacteria can actually help a patient's digestive system heal."

"Wow," Felicia says. "That's amazing! Let's go back to the clubhouse and tell our friends all about it. How do we find our way out?"

"Poop leaves the body through the anus—the hole in your butt," Jae says. "Think of the mouth as the entrance to the digestive system and the anus as the exit. Let's go!"

Gastroenterology

WHOOSH!

"Whoa! That was a quick trip back to the clubhouse," Felicia says.

"And that was an awesome adventure!" Jae says. "We got to see the entire digestive system. We're like junior gastroenterologists now."

"Then how about if we go back to the stomach and fix that cut?"

"That's a job for a real gastroenterologist," Jae says, "but I can tell you how they do it."

"OK," Felicia says. "I'm ready!"

"First, the gastroenterologist makes the patient fall asleep. They do this to keep the patient safe. Next, the doctor fills the stomach with air, like a balloon, so they have lots of space to work."

"Wow," says Felicia. "Then what happens?"

"The gastroenterologist puts an endoscope into the patient's mouth and down the esophagus to the stomach. They use the endoscope to see the inside of the stomach while they work. Then the doctor uses an **endoclip** to close the cut in the stomach. An endoclip is a tool that pinches the cut closed. It stays in the stomach to keep holding the cut closed until it heals."

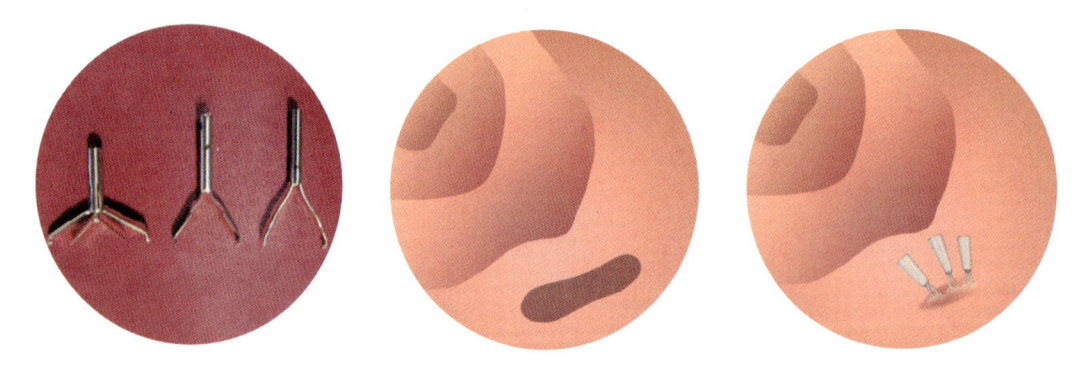

"It's amazing how gastroenterologists can help people," Felicia exclaims. "Thanks for this wild ride."

Gastroenterology 19

"All this talk about digestion is making me hungry," Felicia says, rubbing her grumbling stomach.

"Me too," Jae says.

"I think it's time to eat lunch. What should we eat?"

"Something nutritious," Jae says. "We need to eat foods that keep our bodies healthy."

"Are nutritious foods delicious foods?" Felicia asks.

"They can be," Jae says.

He tells Felicia how fruits and vegetables are filled with lots of nutrients and also taste good.

"We should try to eat them every day," Jae says.

He also tells her that meat, nuts, and beans are healthy foods that give our bodies energy.

"So, what should we eat for lunch?" he asks Felicia.

"How about apples with peanut butter?" she suggests.

"That sounds great!" Jae says.

As Jae and Felicia eat their lunch, they think of all the amazing things their bodies are doing to digest it.

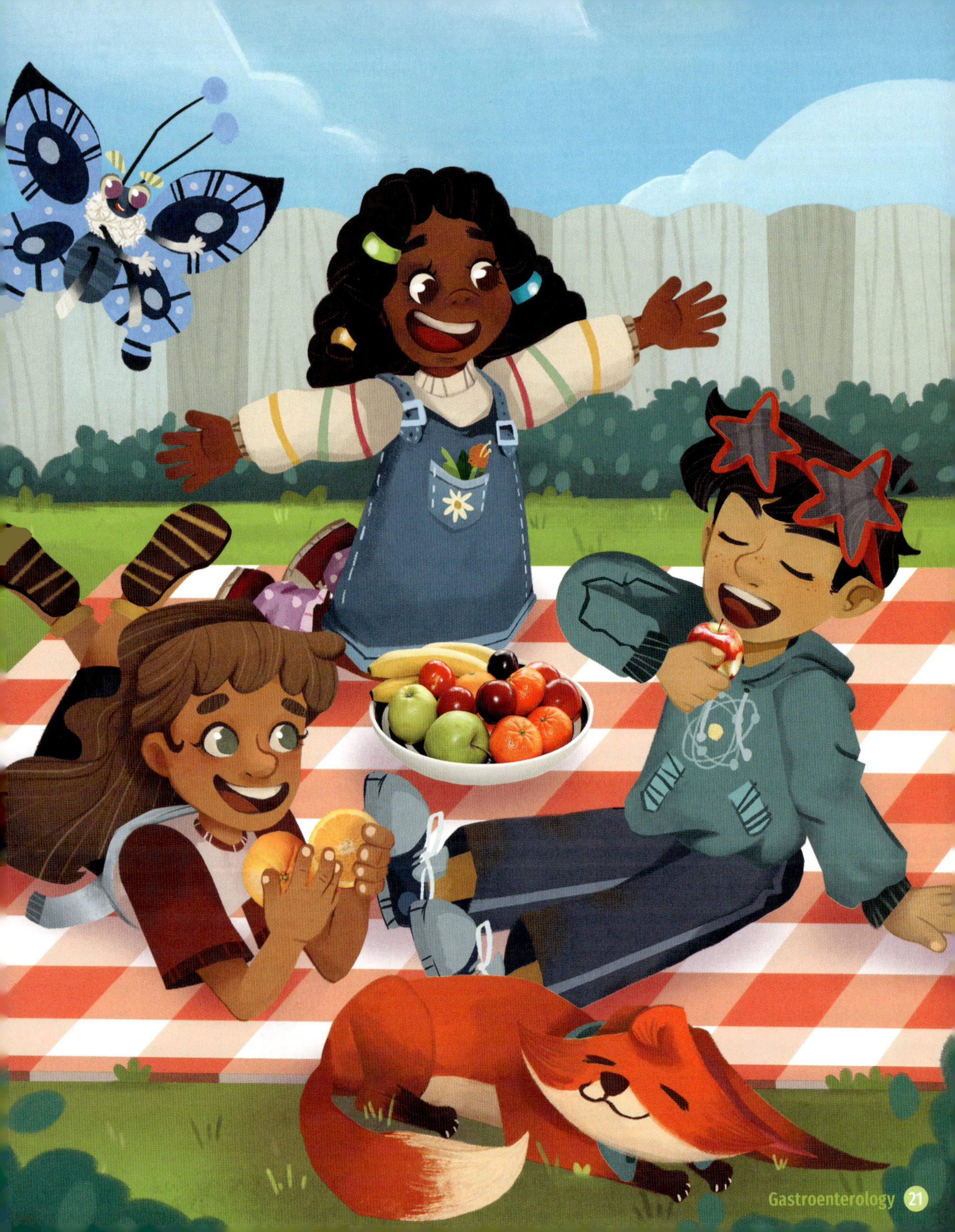

Gastroenterology 21

What is Gastroenterology?

Jae and Felicia gave us a whirlwind tour of the digestive system. Now let's digest all this information and explore some more. Before we tag along with **Dr. Takeshi Saito** for an inside look at how he helps people stay happy and healthy, let's go over some terms that will be helpful in the lab.

Gastroenterology (pronounced GAS-tro-EN-tur-ALL-oh-jee) is the study and medical treatment of the organs in the human body that are responsible for digestion. The esophagus, stomach, small intestine, large intestine, and anus are the organs that are primarily involved in digestion. The organs that produce digestive juices and store the absorbed nutrients are the liver, gallbladder, and pancreas.

Scientists like **Dr. Takeshi Saito** try to prevent things from going wrong with these organs and help us feel better when they do. Let's ask him some questions, and then he'll show us around the hospital!

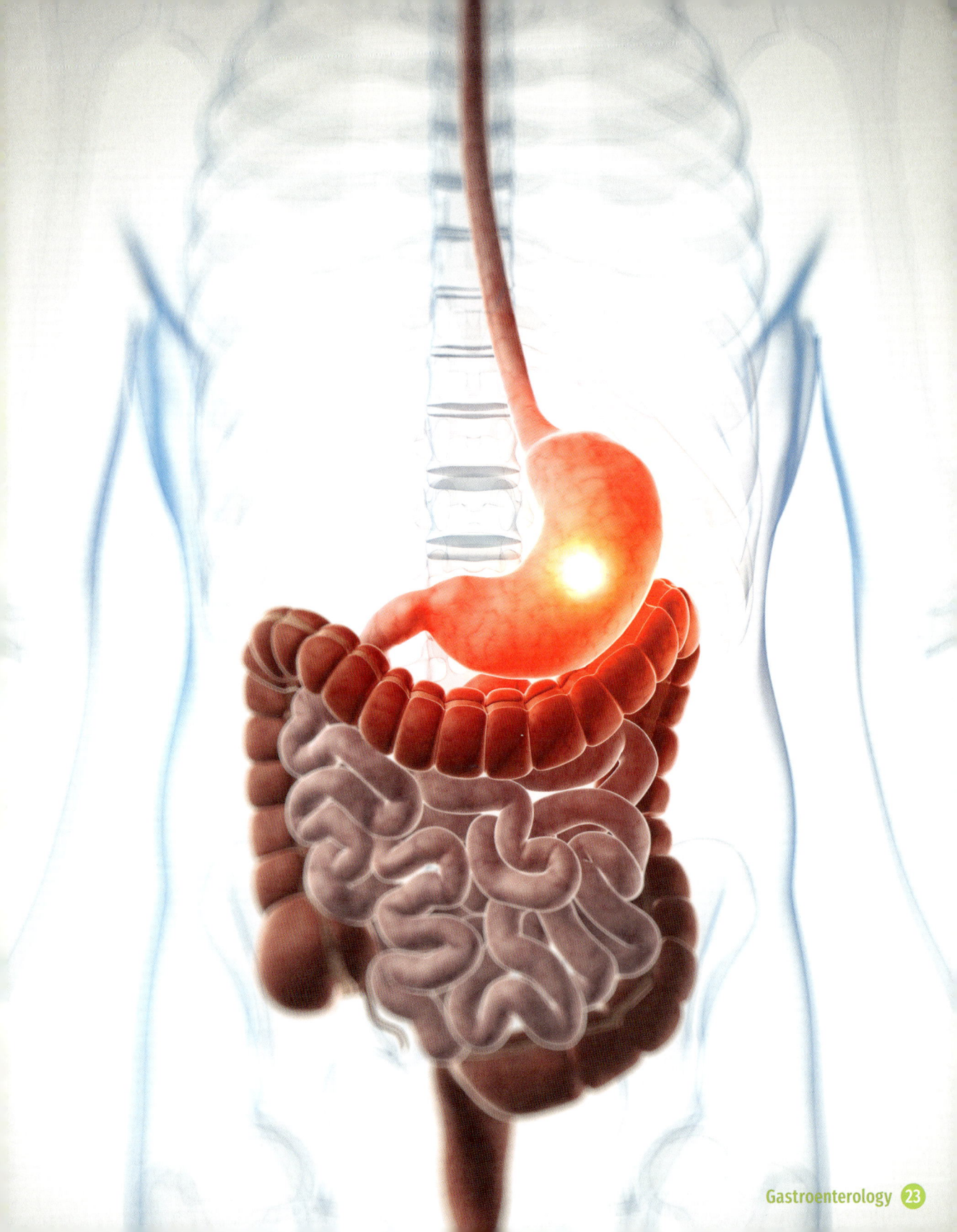

Gastroenterology

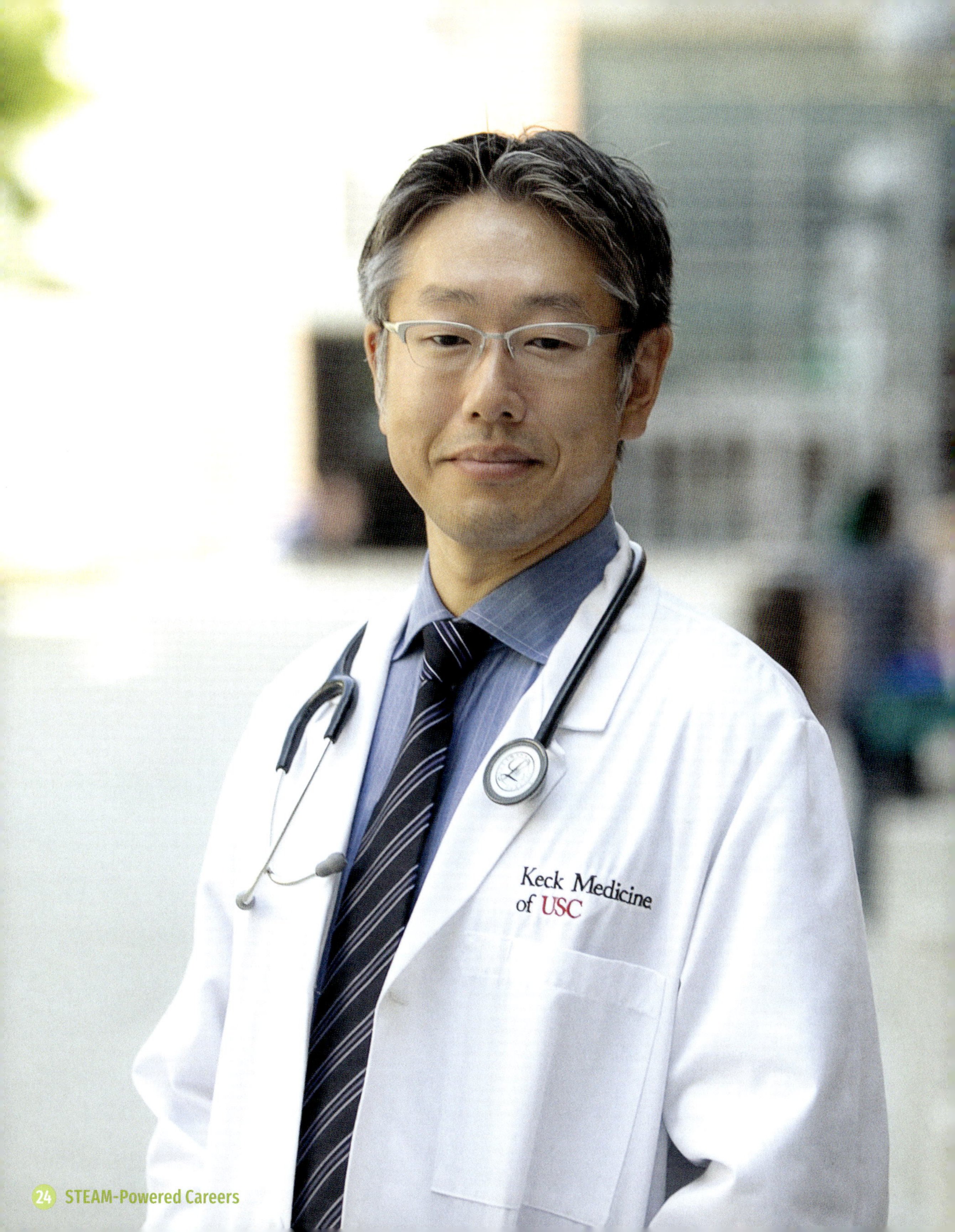
Keck Medicine
of USC

Meet the Scientist
Dr. Takeshi Saito

I have a medical degree with a residency in gastroenterology and a PhD in mircrobiology and virology.

Fun fact #1: I spend about half of my day researching diseases and the other half working with patients.

Fun fact #2: I get to visit many different countries to meet lots of other scientists and see really cool things.

What is your favorite thing about your field?

What is something you would change about the field?

Can you show us your daily routine?

I like learning about how diseases work and how we can cure them.

I wish I could help everyone feel better and make them smile.

I'll show you what my days look like as a doctor and a researcher. Let's head on over!

A Day in the Life (Doctor)

My workday begins when I arrive at a big hospital in Los Angeles, California, called Keck Medicine, at the University of Southern California.

Early in the morning, I look at pictures from an endoscope or from a capsule to diagnose people with different diseases.

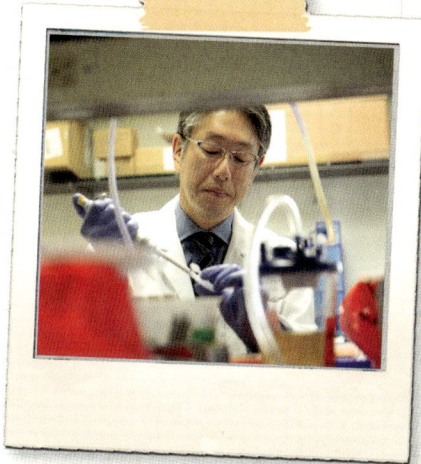

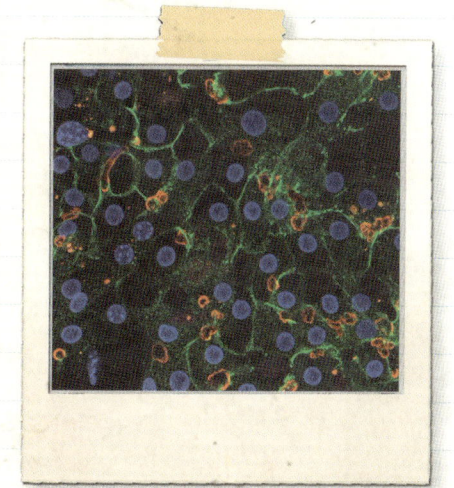

I talk with patients throughout the day, informing them of the diseases they have and telling them about possible treatments.

After my patient appointments, I perform procedures. One procedure I do is an upper endoscopy. I look into a patient's esophagus and stomach to find new things that are making them sick or to fix a problem I have already discovered.

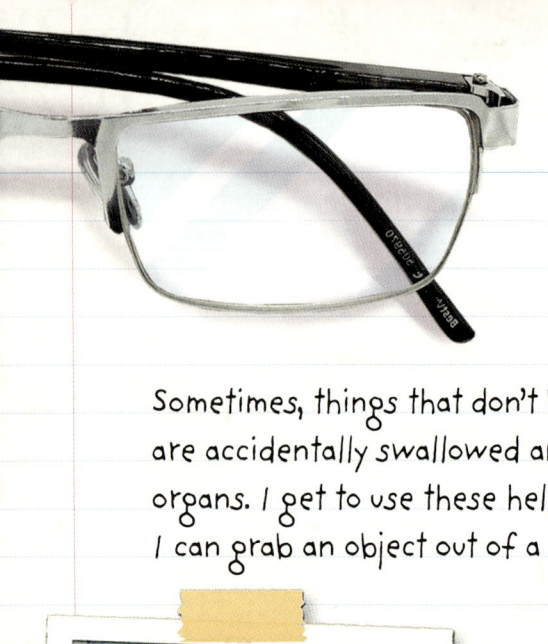

Sometimes, things that don't belong in our bodies are accidentally swallowed and get stuck in our organs. I get to use these helpful tools where I can grab an object out of a patient's organs.

Dinner time @ 7 pm

SAMPLE B

Gastroenterology 27

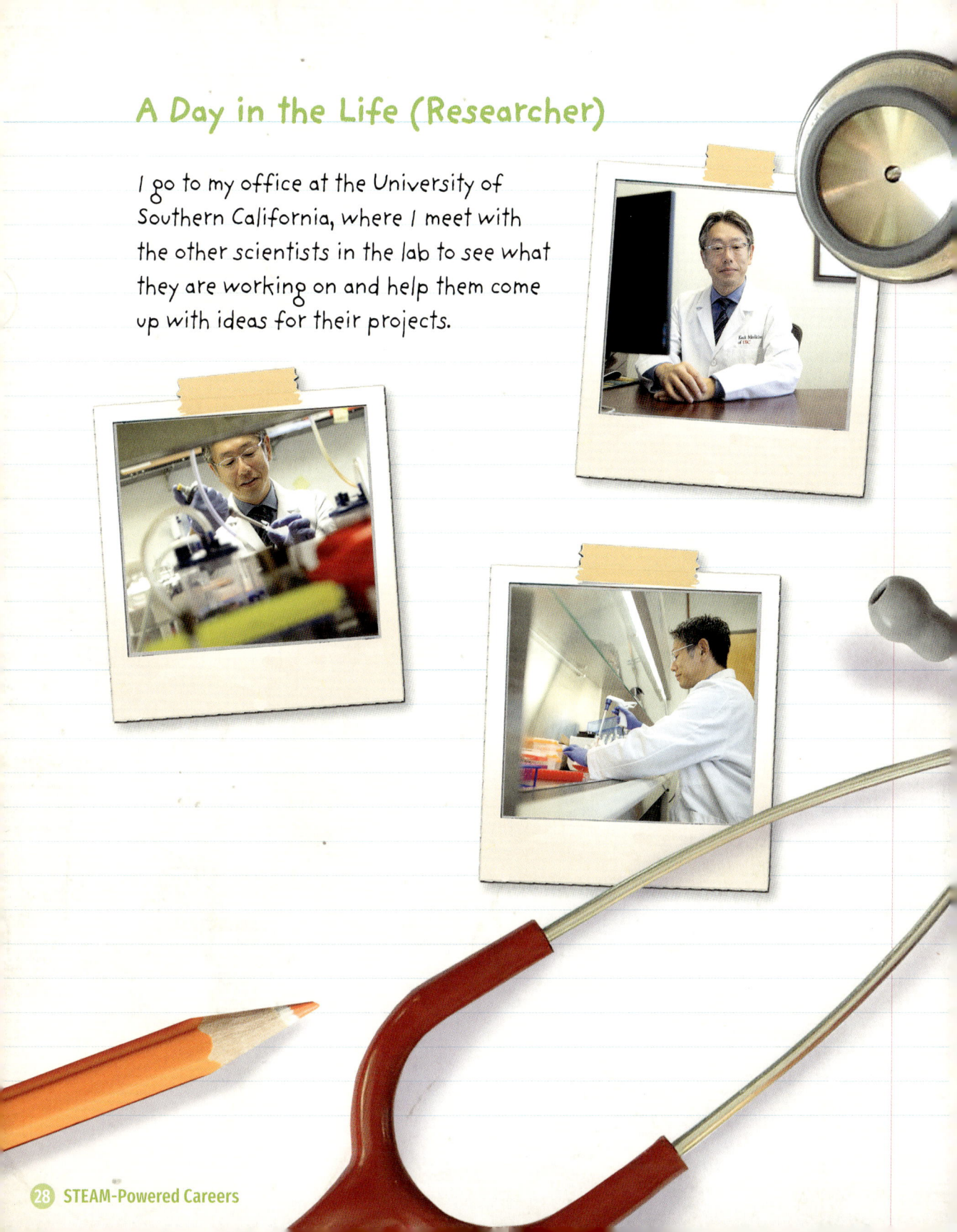

A Day in the Life (Researcher)

I go to my office at the University of Southern California, where I meet with the other scientists in the lab to see what they are working on and help them come up with ideas for their projects.

I use many tools in my work. For example, when conducting my experiments, I use a pipet to mix different liquids. I also use a microscope to look closely at my experiments.

At the end of the day, I return home to be with my family.

Gastroenterology 29

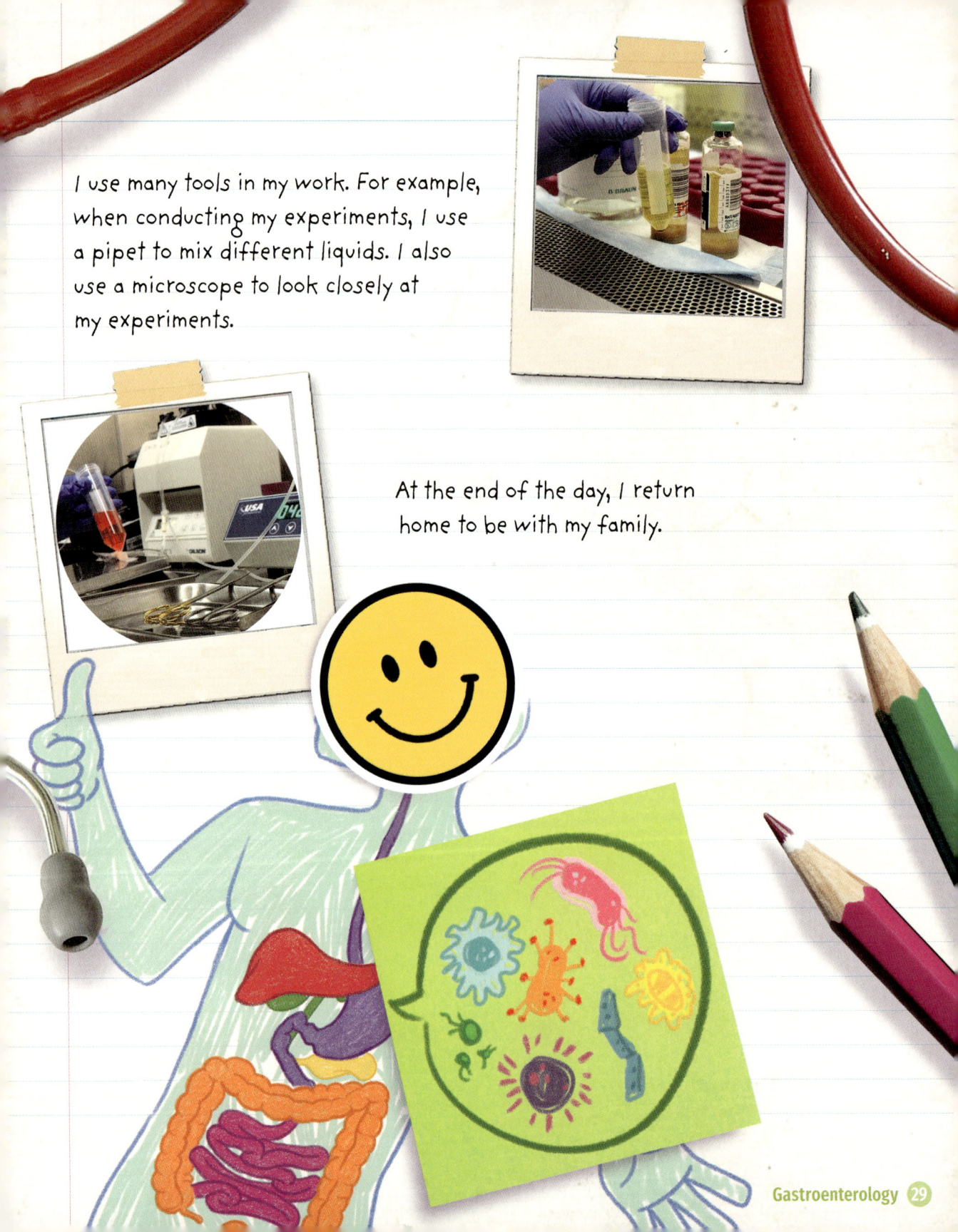

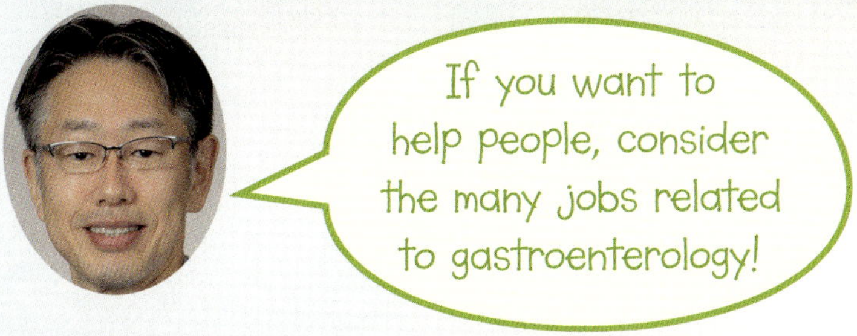

If you want to help people, consider the many jobs related to gastroenterology!

STEAM Careers in Gastroenterology

Gastroenterology is a human body science that plays an important role in our health. Gastroenterologists can become doctors, researchers, and university professors. Some gastroenterologists work in hospitals or private medical offices. Others work in health care clinics or for the government. There are gastroenterologists who specialize in treating children and others who treat adults.

There are many different subjects you can study if you are interested in being a gastroenterologist, including biology, chemistry, human anatomy, genetics, nutrition, physics, statistics, and public health. Gastroenterology is a good career for people who want to help others, are interested in the human body, enjoy doing research, and like working with others.

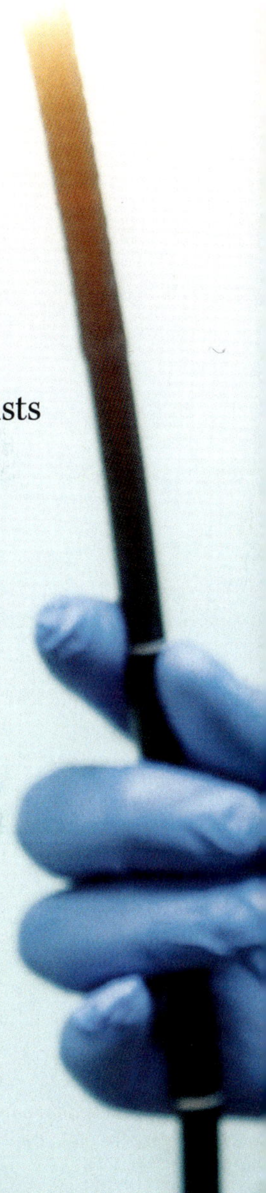

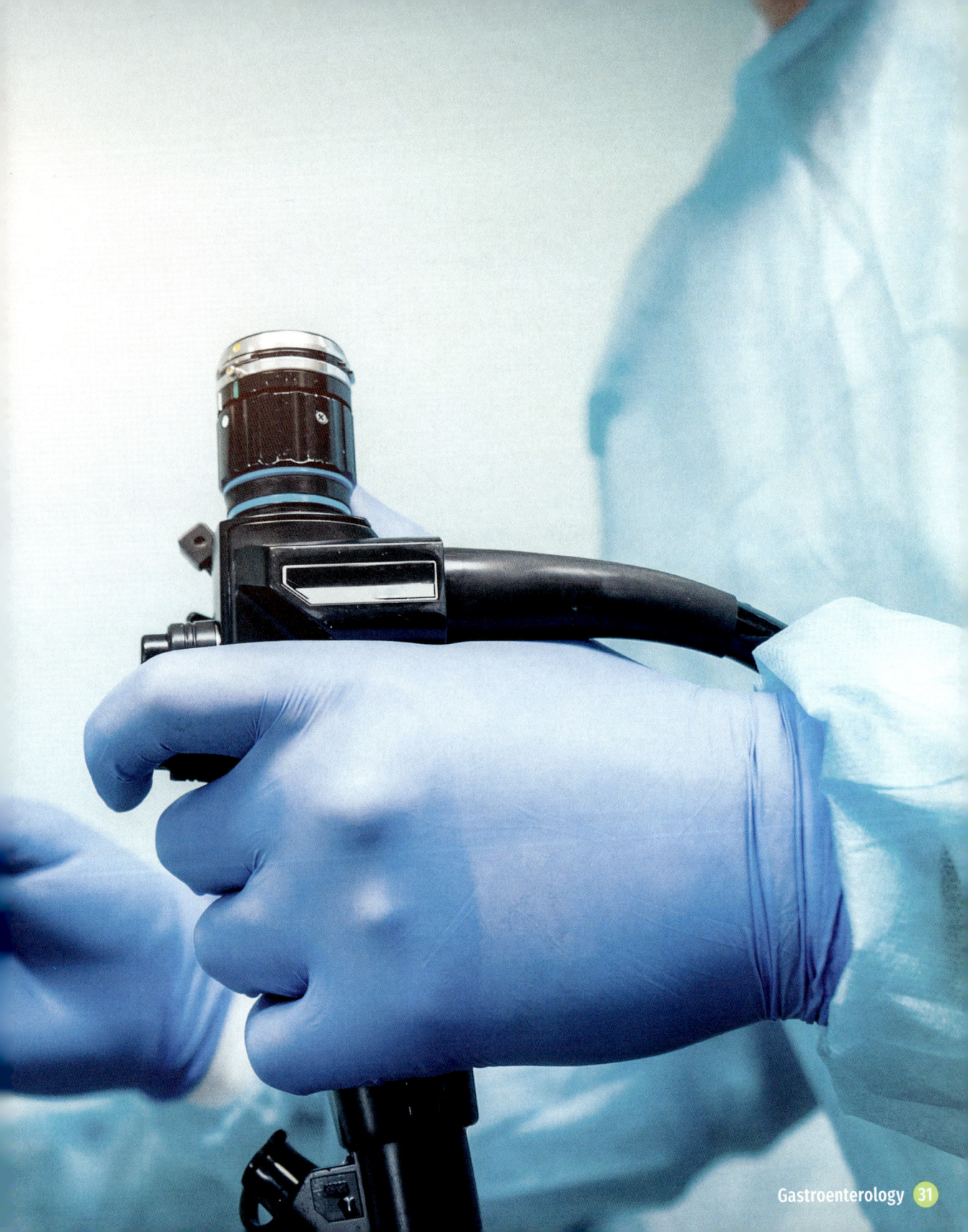

Gastroenterology

The Future of Gastroenterology

In the future, gastroenterologists are going to be finding new cures for diseases and helping people all around the world feel better. New and amazing robots will help gastroenterologists perform surgeries.

Gastroenterologists are going to help people live longer and healthier lives because their patients will have healthy guts and healthy poop!

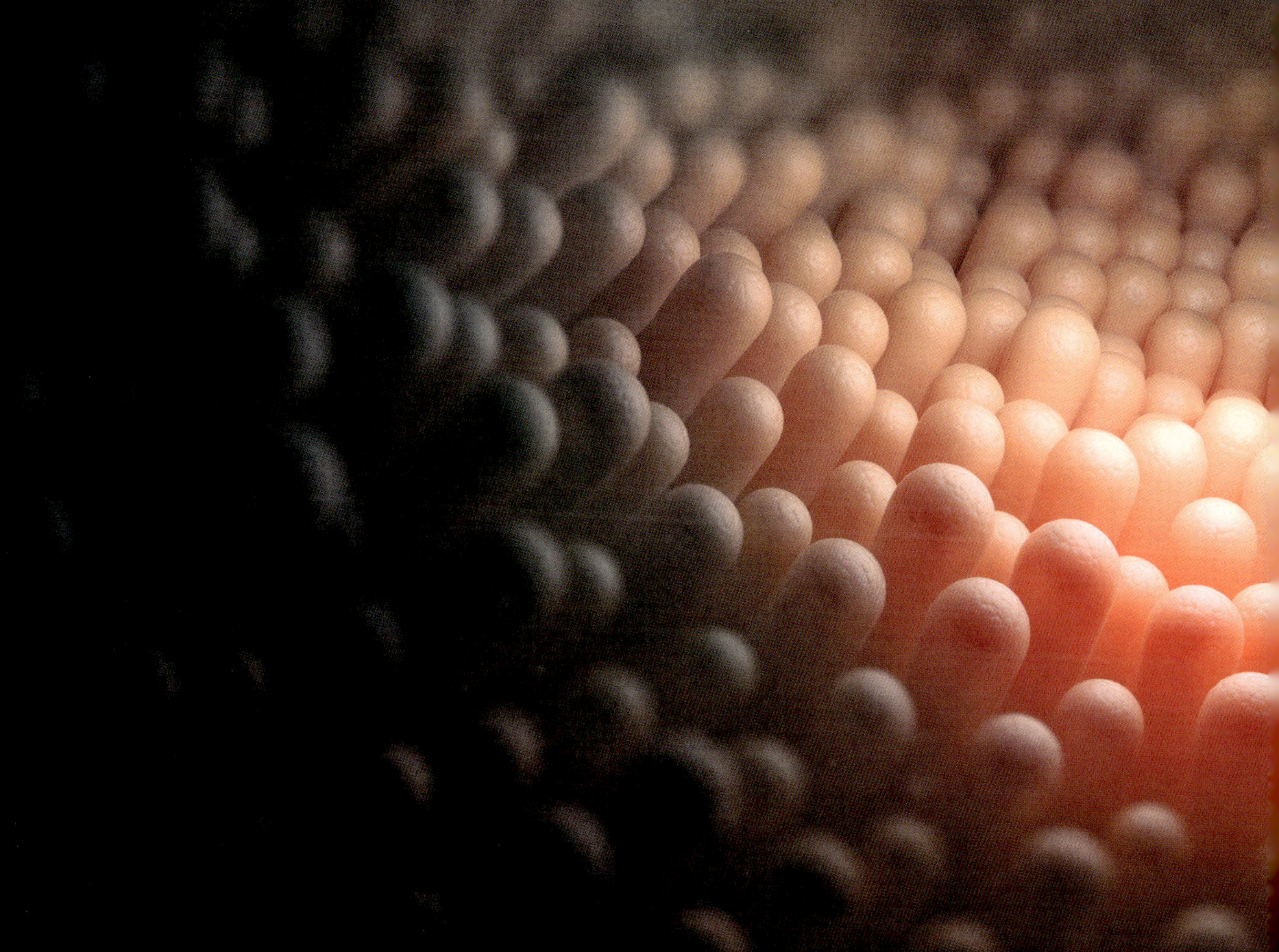

Do You Want to Be a Gastroenterologist?

Spend time learning about the human body. All the different parts of the human body work together to make an awesome human! Watch videos, attend science camps, and participate in your science lessons at school. Practice being a scientist by doing fun experiments!

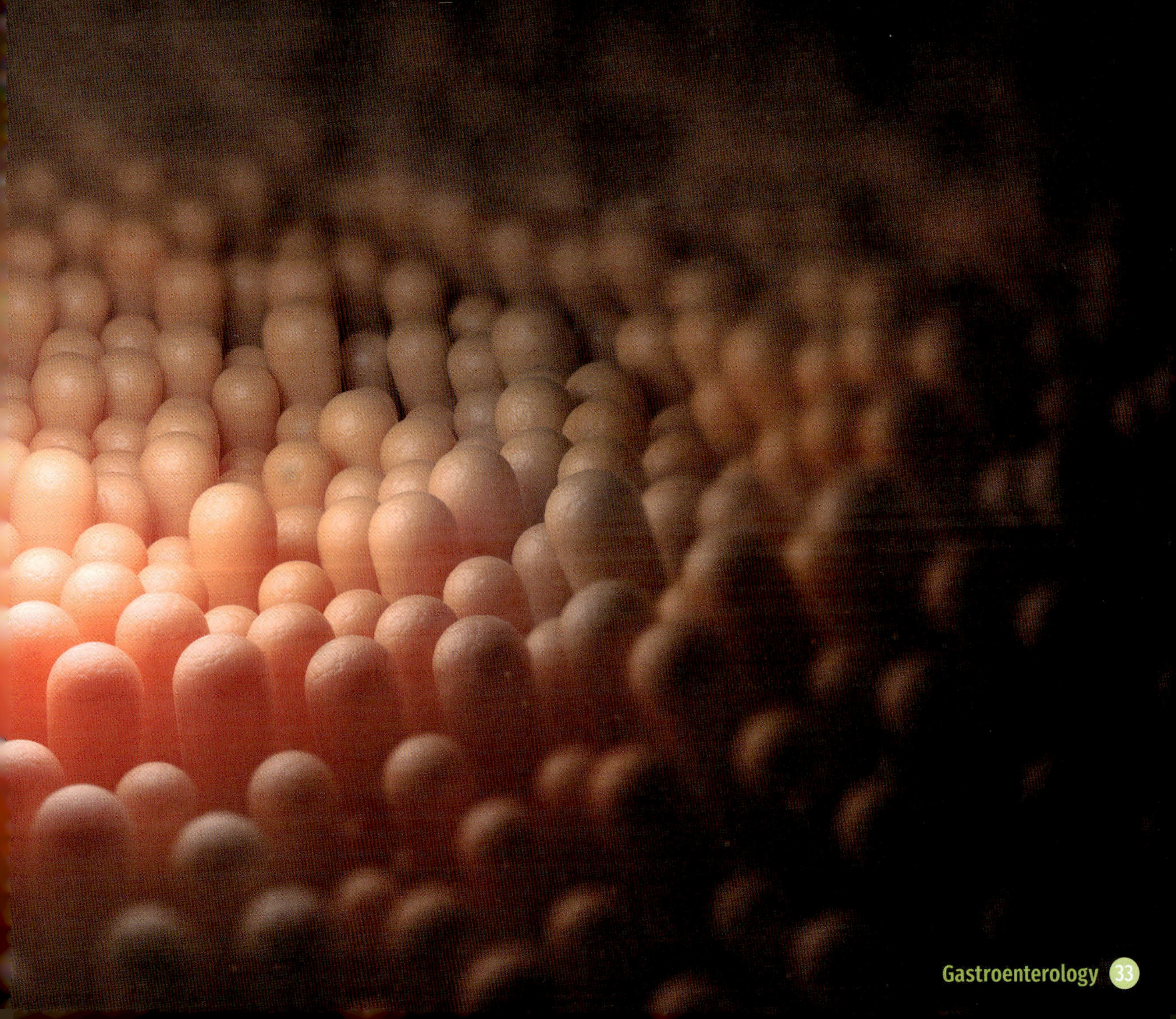

Word List

bacteria: very small living things that are found almost everywhere on Earth, even in the human body

digestion: the breaking down of food to be turned into *nutrients* and energy for the body

- **chemical digestion:** when *stomach acid*, *bacteria*, and *enzymes* break down food
- **mechanical digestion:** when food is smashed into tiny pieces by chewing and from the movement of the organs

endoclip: a metal device used to close cuts in organs

endoscope: a camera that takes pictures of the inside of your body

endoscopy: the medical procedure that uses a camera at the end of a long tube to look at the gastrointestinal organs

enzymes: special chemicals in the human body that break down food during *chemical digestion*

esophagus: a muscular tube that connects the throat to the stomach

nutrients: substances that living things need in order to grow

stomach acid: a liquid in the stomach that is full of good *bacteria* and *enzymes* that help break down food during *digestion*

villi: (singular, villus) finger-like structures in the small intestine that absorb *nutrients* from food

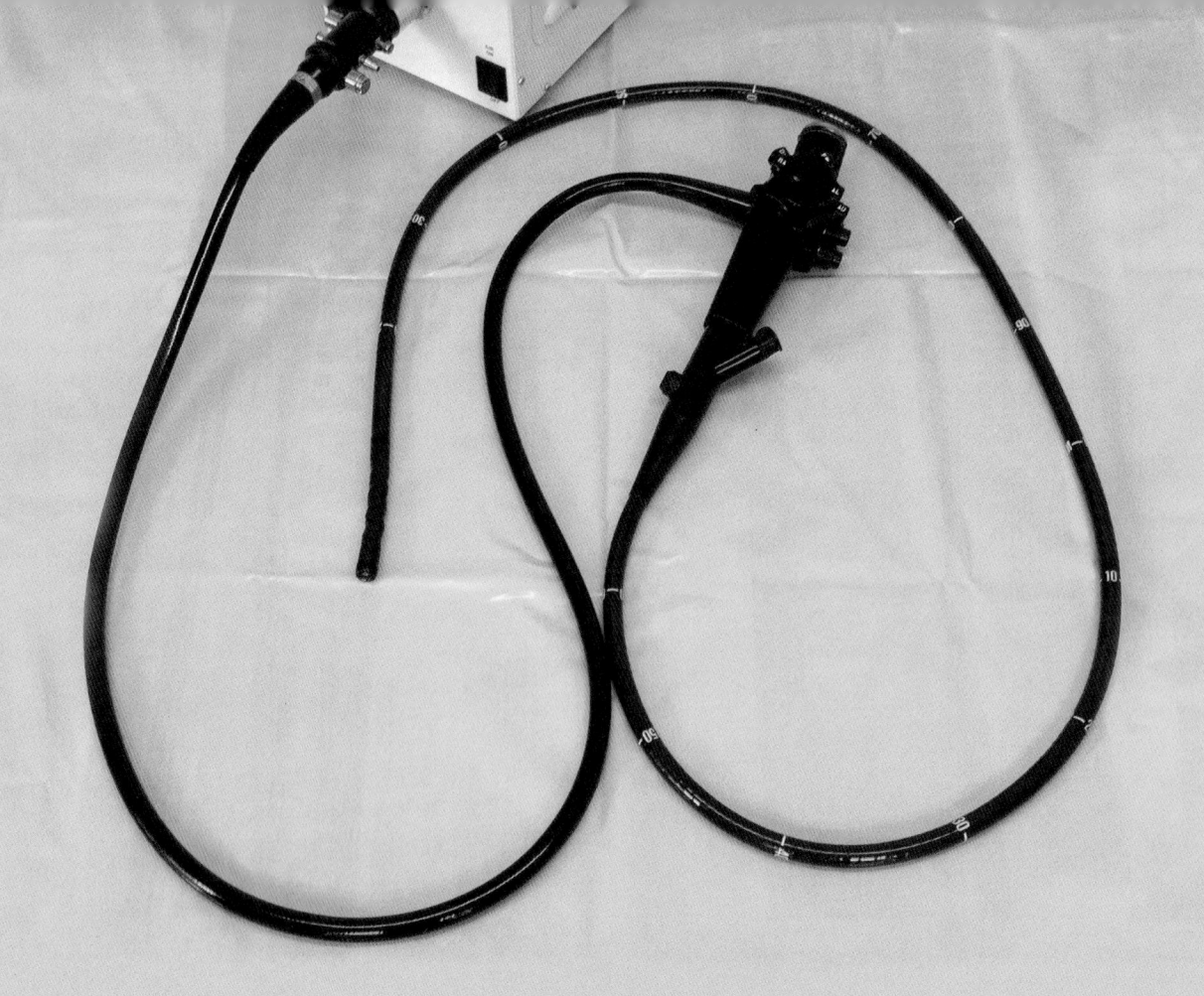

Gastroenterology Resources

Check out these books:

The Quest to Digest by Mary K. Corcoran

Curious George Goes to the Hospital by Margret and H. A. Rey.

Upper endoscopy guide for kids:
   https://youtu.be/1PrlgWWiJGchttps://youtu.be/1PrlgWWiJGc

Upper endoscopy details:
   https://www.aboutkidshealth.ca
   /Article?contentid=2472&language=English

Acknowledgments

University of Southern California, Joint Educational Project STEM Education Programs and USC Norris Comprehensive Cancer Center

Brooke Allyn is a high school science teacher in Los Angeles, California. Her mission is to deliver a high-quality STEAM education to traditionally underserved students. When she's not teaching, she loves spending time with her many animals.

Dr. Takeshi Saito lives in Los Angeles, California. He is originally from Japan, where he completed his medical and science education. He works at the University of Southern California (USC) Keck School of Medicine as an associate professor of medicine, molecular microbiology, immunology, and pathology. At USC he functions as a physician-scientist who sees patients and runs a basic research laboratory investigating the molecular mechanisms of liver diseases.

Janet Pagliuca is a Venezuelan children's book illustrator and designer who found her passion for books at a young age. She graduated from Savannah College of Art and Design with a BFA in Communication Arts. Her inspiration comes from her personal experience growing up in two different cultures, which taught her about self-identity. She hopes to keep creating children's books that depict the importance of cultural awareness and diversity.

Explore the Complete

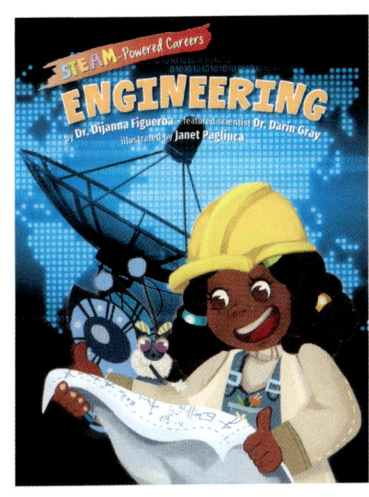

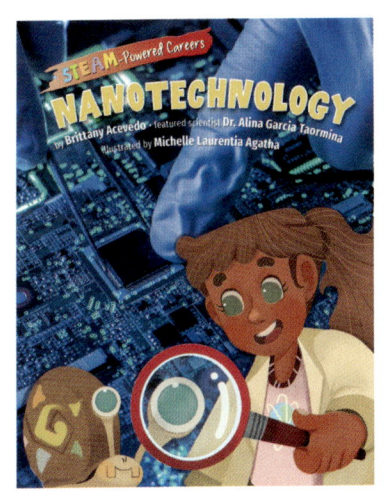

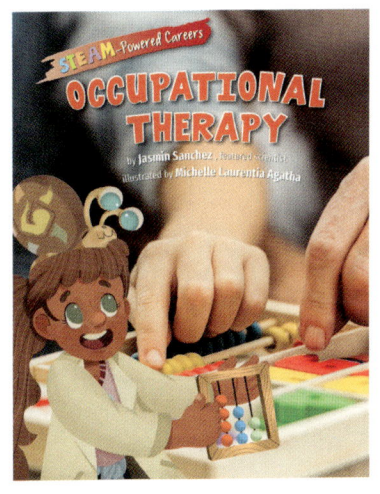

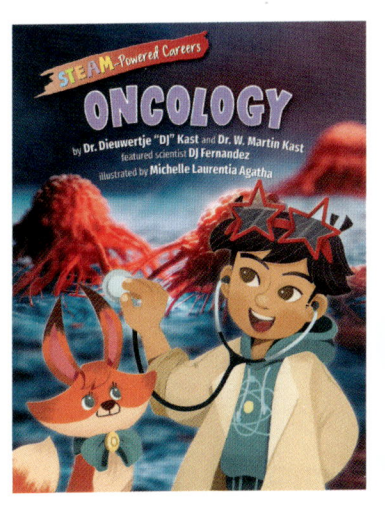

STEAM-Powered Careers: DATA SCIENCE
by Stacey Finley, featured scientist
illustrated by Michelle Laurentia Agatha

STEAM-Powered Careers: ENGINEERING
by Dr. Dijanna Figueroa, featured scientist Dr. Darin Gray
illustrated by Janet Pagliuca

STEAM-Powered Careers: NANOTECHNOLOGY
by Brittany Acevedo, featured scientist Dr. Alina Garcia Taormina
illustrated by Michelle Laurentia Agatha

STEAM-Powered Careers: OCCUPATIONAL THERAPY
by Jasmin Sanchez, featured scientist
illustrated by Michelle Laurentia Agatha

STEAM-Powered Careers: ONCOLOGY
by Dr. Dieuwertje "DJ" Kast and Dr. W. Martin Kast
featured scientist DJ Fernandez
illustrated by Michelle Laurentia Agatha

STEAM-Powered Careers Series!

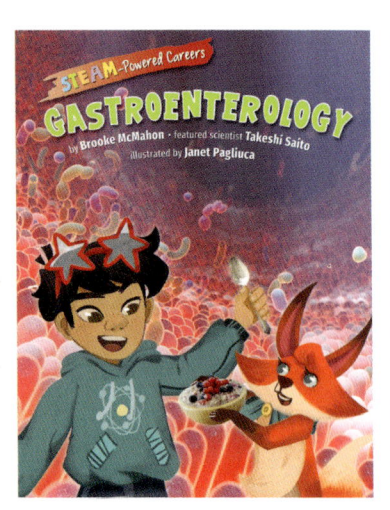

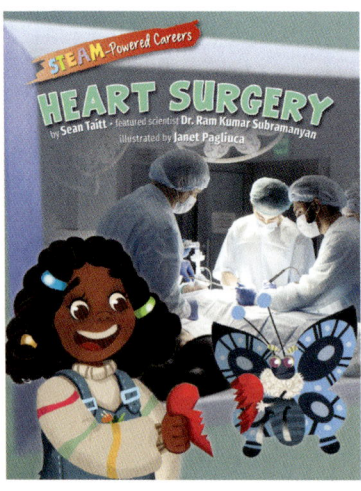

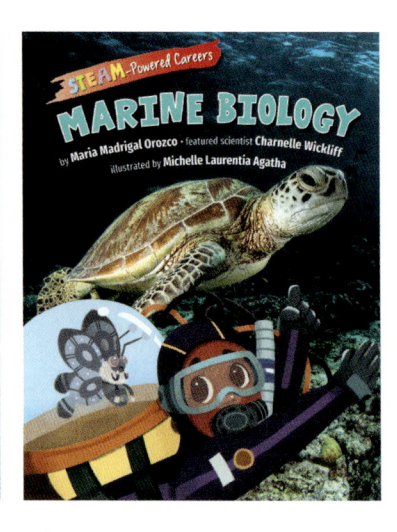

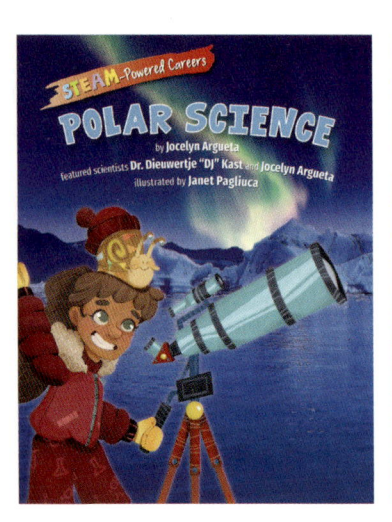

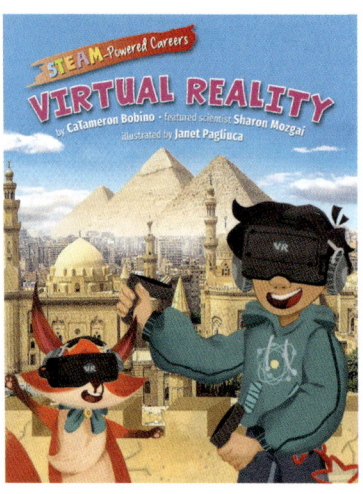

GASTROENTEROLOGY
by Brooke McMahon • featured scientist Takeshi Saito
illustrated by Janet Pagliuca

HEART SURGERY
by Sean Taitt • featured scientist Dr. Ram Kumar Subramanyan
illustrated by Janet Pagliuca

MARINE BIOLOGY
by Maria Madrigal Orozco • featured scientist Charnelle Wickliff
illustrated by Michelle Laurentia Agatha

POLAR SCIENCE
by Jocelyn Argueta
featured scientists Dr. Dieuwertje "DJ" Kast and Jocelyn Argueta
illustrated by Janet Pagliuca

VIRTUAL REALITY
by CaTameron Bobino • featured scientist Sharon Mozgai
illustrated by Janet Pagliuca

Photo credits

Cover iStock.com/Marcin Klapczynski **6–7** iStock.com/Oleksandr Hruts; © Leonello Calvetti | Dreamstime.com; iStock.com/SciePro; iStock.com/Viktoriya Kabanova **8** coddie/Depositphotos.com **10** iStock.com/Dmytro Lastovych **12–13** © Maubuk77 | Dreamstime.com **14–15** maxxyustas/Depositphotos.com; iStock.com/Jakub Rupa **16–17** BRRT/Pixabay **18–19** 2014 Shou-jiang Tang, "Endoscopic Management of Gastrocutaneous Fistula Using Clipping, Suturing, and Plugging Methods," Figure 2 (A). Endoscopic through-the-scope (TTS) clip placement Published by Elsevier Gmb, CC BY 3.0, https://doi.org/10.1016/j.vjgien.2014.03.001 **20–21** Desertrose7/Pixabay; Erbs55/Pixabay **23** iStock.com/SciePro **24** Ricardo Carrasco III, Keck Medicine, USC **26–27** Ricardo Carrasco III, Keck Medicine, USC; Dr. Takeshi Saito, Keck Medicine; Ricardo Carrasco III, Keck Medicine, USC; © Alexei Averianov | Dreamstime.com; © Natalia Sokolova | Dreamstime.com **28–29** Ricardo Carrasco III, Keck Medicine, USC; Dr. Takeshi Saito, Keck Medicine; Colin, CC BY-SA 4.0, via Wikimedia Commons **30–31** Romaset/Depositphotos.com **32–33** © Siarhei Yurchanka | Dreamstime.com **34–35** Chrispo/Depositphotos.com **36–37** inside1703/Depositphotos.com; Liam Mayr; Ricardo Carrasco III, Keck Medicine, USC; photo courtesy of Janet Pagliuca **40** © Senkumar Alfred | Dreamstime.com

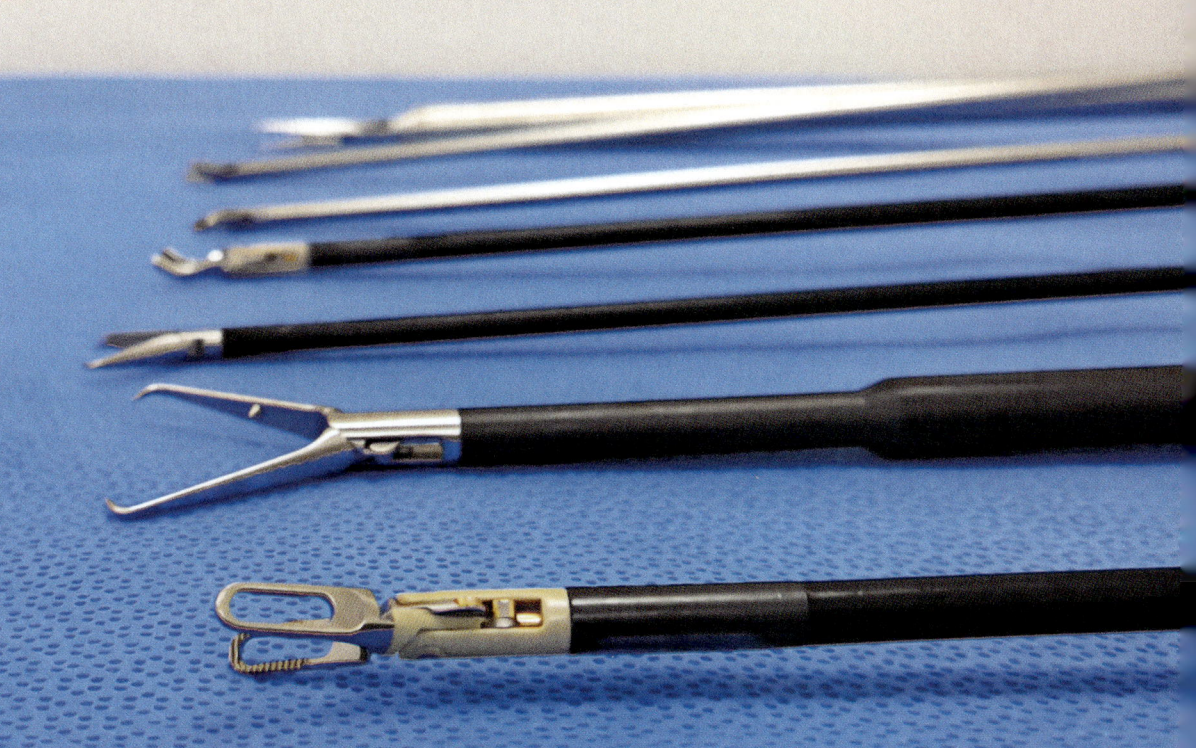